The Shanghai Maths Project

For the English National Curriculum

一课一练

Year 6 Learning

Collins

William Collins' dream of knowledge for all began with the publication of his first book in 1819.

A self-educated mill worker, he not only enriched millions of lives, but also founded a flourishing publishing house. Today, staying true to this spirit, Collins books are packed with inspiration, innovation and practical expertise. They place you at the centre of a world of possibility and give you exactly what you need to explore it.

Collins. Freedom to teach.

Published by Collins
An imprint of HarperCollins*Publishers*
The News Building
1 London Bridge Street
London
SE1 9GF

Browse the complete Collins catalogue at
www.collins.co.uk

© HarperCollins*Publishers* Limited 2018

10 9 8 7 6 5 4 3 2 1

978-0-00-822600-8

Learning Books Series Editor: Amanda Simpson

Practice Books Series Editor: Professor Lianghuo Fan

Authors: David Bird, Linda Glithro, Paul Hodge, Jo Lees, Richard Perring and Paul Wrangles

British Library Cataloguing in Publication Data

A catalogue record for this publication is available from the British Library.

Publishing Manager: Fiona McGlade and Lizzie Catford
In-house Editor: Mike Appleton
In-house Editorial Assistant: August Stevens
Project Manager: Karen Williams
Copy Editor: Karen Williams
Proofreaders: Catherine Dakin and Steven Matchett
Cover design: Kevin Robbins and East China Normal University Press Ltd
Cover artwork: Daniela Geremia
Internal design: Amparo Barrera
Typesetting: Ken Vail Graphic Design Ltd
Illustrations: Ken Vail Graphic Design Ltd and QBS
Production: Sarah Burke

Printed and bound in Latvia

Writing decimals

A decimal number is one that includes tenths, hundredths, thousandths or a combination of these.

$$5.371 = 5 + \frac{3}{10} + \frac{7}{100} + \frac{1}{1000}$$

Usually, the same number can be written as both a decimal and as a fraction.

Whether it's written as a decimal or a fraction doesn't change the value. It is in the same position on the number line.

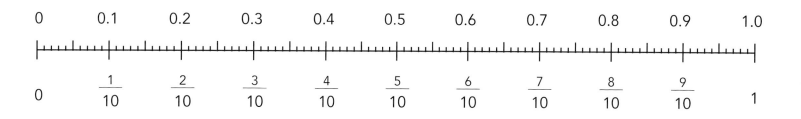

When the decimal part of a number ends in a zero, the zero can be removed without changing the value of the number.

These all look different, but they have the same value.

$$7.8100 = 7.81 = 7.810\,000\,000\,000\,000\,000\,000\,000\,000\,000\,000\,000$$

Units

Kilo means 1000 of a unit, so a kilogram is 1000 grams and a kilometre is 1000 metres.

Centi means $\frac{1}{100}$ of a unit, so a centimetre is $\frac{1}{100}$ of a metre and a centilitre is $\frac{1}{100}$ of a litre.

Milli means $\frac{1}{1000}$ of a unit, so a millimetre is $\frac{1}{1000}$ of a metre and a milligram is $\frac{1}{1000}$ of a gram.

Comparing decimals

Which is greater: 0.25 or 0.7?

A place value chart can help.

100	10	1	.	0.1	0.01
		0	.	②	5

100	10	1	.	0.1	0.01
		0	.	⑦	

Writing the decimals as fractions can help.

$$0.25 = \frac{2}{10} + \frac{5}{100} \qquad 0.7 = \frac{7}{10}$$

There are two tenths in 0.25 and there are seven tenths in 0.7. So, 0.7 is greater than 0.25.

Adding and subtracting decimals

Adding and subtracting decimals using the column method follows the same rules as adding and subtracting whole numbers.

Just remember to line up the decimal points one above the other.

$1.369 + 7.58 = \boxed{}$

```
    1 . 3  6  9
+   7 . 5  8
  _____
    8 . 9  4  9
           1
```

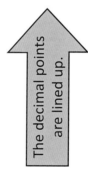

The decimal points are lined up.

$3.142 - 1.071 = \boxed{}$

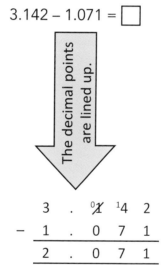

The decimal points are lined up.

```
    3 . ⁰3̸ ¹4  2
-   1 . 0   7  1
  _____
    2 . 0   7  1
```

Rounding decimals

Rounding is a way of approximating decimal numbers.

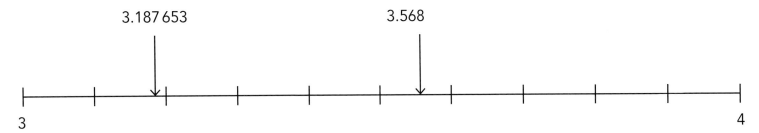

	3.187 653 is **between** 3.1 and 3.2	3.568 is **between** 3.5 and 3.6
Rounded to the nearest whole tenth	3.2	3.6
Rounded up to the next whole tenth	3.2	3.6
Rounded down to the previous whole tenth	3.1	3.5

Parts of a circle

diameter = 2 × radius

$$d = 2r$$

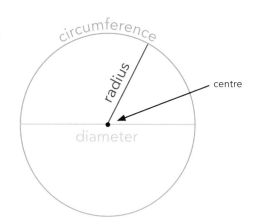

Angles on a straight line

A straight angle contains two right angles, so it is 180°.

Angles on a straight line that combine to make a straight angle will have a sum of 180°.

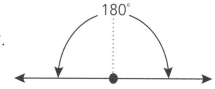

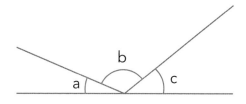

$\angle a + \angle b + \angle c = 180°$

Multiplying and dividing decimals by 10, 100 or 1000

The digits move across the decimal point.

100	10	1	.	0.1	0.01
		3	.	1	
	3	1	.		

Digits move one place to the left when multiplied by 10.

$3.1 \times 10 = 31$

Digits move two places to the right when divided by 100.

$5.2 \div 100 = 0.052$

100	10	1	.	0.1	0.01	0.001
		5	.	2		
		0	.	0	5	2

Remember to fill any gaps to the left with zeros.

It can be helpful to think of a place value slider when multiplying and dividing by 10, 100 or 1000.

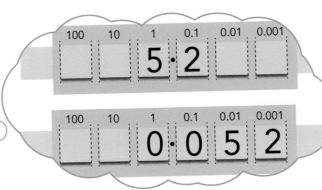

Multiplying a whole number by a decimal

Multiplying a whole number by a decimal can be carried out by converting to whole numbers.

The product of 1.6 × 23 is one tenth of the product of 16 × 23. So, instead of calculating with the decimal, we can multiply with whole numbers and then divide the product by 10.

```
      2  3                          2  3
×     1 . 6      × 10      ×        1  6
  ───────────   ────────>     ──────────
  1   3 . 8                    1   3  8
      2  3 . 0    ÷ 10             2  3  0
  ───────────   <────────     ──────────
  3   6 . 8                    3   6  8
```

16 × 23 = 368

So, 1.6 × 23 = (the product of 16 × 23) ÷ 10.

Dividing a decimal by a whole number

Dividing a decimal by a whole number can be carried out by converting to whole numbers.

In the same way, the quotient of 4.7 ÷ 5 is one tenth of the quotient of 47 ÷ 5.

```
      0  9 . 4
  5 ) 4  7 . 0
      4  5
      ─────
         2 . 0
         2 . 0
      ─────
         0 . 0
```

$$47 \div 5 = 9.4$$

$$\boxed{4 \; 7} \div 5 = \boxed{9.4}$$

Since 4.7 was multiplied by 10 to find the quotient, the quotient itself must now be divided by 10 to show the quotient of the original dividend.

$$9.4 \div 10 = 0.94$$

$$\boxed{9.4} \div 10 = \boxed{0.94}$$

Approximating

Rounding a decimal gives an approximation of the number.

A number line can help to find approximations.

Expressing 5.162 to the nearest hundredth:
- We know that it is between 5.16 (5 and 16 hundredths) and 5.17 (5 and 17 hundredths).
- It is closer to 5.16, and so 5.162 rounded to the nearest hundredth is 5.16.

Expressing 5.162 to the nearest tenth:
- It is between 5.1 (5 and 1 tenth) and 5.2 (5 and 2 tenths).
- It is closer to 5.2, and so 5.162 rounded to the nearest tenth is 5.2.

Rules

Unknown values ('variables') in equations are represented by letters.

The value of r is not known: $4r + 6 = 22$ (but can be worked out).

The letter r represents the number 4.

Letters can represent numbers in sequences.

4, 7, 10, Z, 16, …   The letter Z represents an unknown value. We can work out that Z represents 13.

When a number and a letter are written together, this means they are multiplied.

The number goes first: $8m$ means 8 multiplied by m.

$0.6(2w)$ means 0.6 multiplied by the product of $2 \times w$.

Sequences

A term-to-term rule can create a sequence.

3, 8, 13, 18, 23, …

The term-to-term rule is + 5.

An nth term can create a sequence.

4, 9, 14, 19, 24, …

The nth term for this sequence is $5n - 1$ where n is the position of the number in the sequence.

Bar models can be used to represent equations

$3x + 7 = 22$

22			
x	x	x	7

$4y - 8 = 20$

y	y	y	y
20		8	

Expansion of brackets

$a(b + c) = a \times b + a \times c = ab + ac$

An array can be used to represent the expansion.

	a
b	ab
c	ac

$a(b + c) = a \times b + a \times c = ab + ac$

Lines

Intersecting lines

Two or more lines that meet or cross at a point are called intersecting lines.

The point where the lines intersect is called the point of intersection.

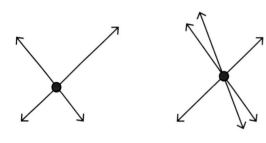

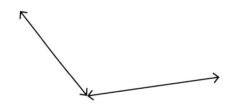

There is no limit to the number of lines that can share a point of intersection.

Perpendicular lines

Lines that meet or cross at a 90° angle are called perpendicular lines.

Unlike parallel lines that never touch, perpendicular lines must intersect.

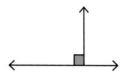

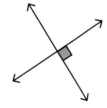

A small box symbol is used to indicate 'at right angles'.

Parallel lines

Lines that are always the same distance apart and never touch are called parallel lines.

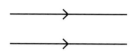

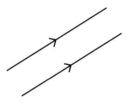

Small arrows are used to show lines that are parallel to each other.

Parallelograms

A parallelogram is a four-sided shape with parallel opposite sides that are equal in length.

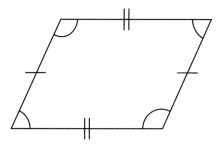

| and || show equal opposite sides.
Opposite angles are also equal.

A rectangle is a special form of a parallelogram with four right angles.

rectangle   square

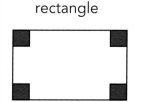 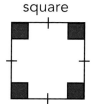

A square is a special parallelogram with four right angles and four equal sides. A square is also a special rectangle.

Any side of a parallelogram can be its base.

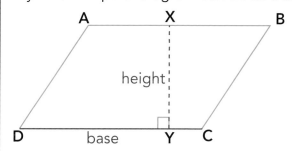

The height is the perpendicular distance from the base to the opposite side.
We mark the height with a dashed line to avoid confusion with the sides of the shape.

Area of a parallelogram = base × height.

$$A = b \times h$$

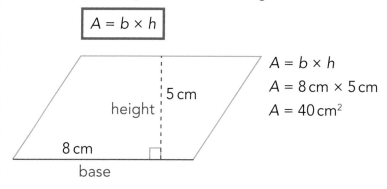

$A = b \times h$
$A = 8\,cm \times 5\,cm$
$A = 40\,cm^2$

Triangles

Any side of a triangle can be its base.

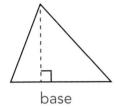 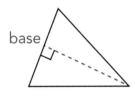 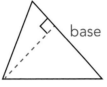

base

base

base

base

The height is the perpendicular distance from the base to the opposite vertex.

Area of a triangle

Area of a triangle = half × base × height.

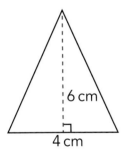

6 cm

4 cm

$A = \frac{1}{2} \times b \times h$

$A = \frac{1}{2} \times 6\,cm \times 4\,cm$

$A = 12\,cm^2$

$$A = \frac{1}{2} \times b \times h$$

$A = \frac{1}{2} \times b \times h$

$A = \frac{1}{2} \times 10\,cm \times 4\,cm$

$A = 20\,cm^2$

4 cm

height

10 cm

base

Simplifying fractions using common factors

We can find the simplest form of a fraction by following these steps.

$$\frac{18}{24}$$

Step 1: Find the factors of the numerator and denominator.

factors of 18	factors of 24

①②③⑥ 9 and 18   ①②③ 4 ⑥ 8, 12 and 24

Step 2: Identify common factors.
The highest common factor is **6**.

Step 3: Divide both the numerator and denominator by the **highest common factor**.
The fraction is now in its **simplest form**.

$$\frac{18}{24} = \frac{(3 \times 6)}{(4 \times 6)}$$

$$\frac{18}{24} = \frac{3}{4}$$

Finding the lowest common denominator

When two fractions have a common denominator (the same denominator) they can be compared easily, and operations can be carried out.

To make the denominators the same, a common multiple must be found.

$$\frac{3}{4} + \frac{5}{6} = \square$$

Multiples of 4:
4, 8, ⑫ 16, 20, ㉔ 28, 32, ㊱ 40, ...

Multiples of 6:
6, ⑫ 18, ㉔ 30, ㊱ 42, ...

12, 24 and 36 are common multiples of 4 and 6.

Use the lowest common multiple as the new denominator for both fractions. The calculation is now straightforward:

$$\frac{3}{4} + \frac{5}{6} = \frac{9}{12} + \frac{10}{12}$$ (both fractions can be converted to twelfths and this is the **lowest common denominator**)

Adding and subtracting fractions

We can add and subtract fractions by following these steps.

Step 1: Find the lowest common denominator and convert both fractions so they have the same denominator.

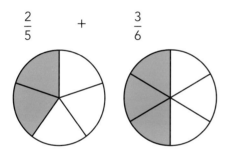

$$\frac{2}{5} \quad + \quad \frac{3}{6}$$

Step 2: Add the numerators (or subtract if it is a subtraction). This will tell you how many of the denominator you have now.

$$\frac{2}{5} + \frac{3}{6} = \frac{12}{30} + \frac{15}{30}$$

$$\frac{12}{30} \quad + \quad \frac{15}{30} \quad = \quad \frac{27}{30}$$

Step 3: Simplify if possible. Is there a common factor? What is the lowest common factor?

$$\frac{27}{30} = \frac{(9 \times 3)}{(10 \times 3)} = \frac{9}{10}$$

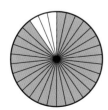

Multiplying a fraction by a fraction

Arrays help us to visualise what happens when a fraction is multiplied by a fraction. Can you see that:

$\frac{1}{3} \times \frac{1}{4} = \frac{1}{12}$ is the same as $\frac{1}{3}$ of $\frac{1}{4}$?

$\frac{2}{3} \times \frac{1}{4} = \frac{2}{12}$ is the same as $\frac{2}{3}$ of $\frac{1}{4}$?

$\frac{2}{3} \times \frac{3}{4} = \frac{6}{12}$ is the same as $\frac{2}{3}$ of $\frac{3}{4}$?

$\frac{1}{3} \times \frac{3}{4} = \frac{3}{12}$ is the same as $\frac{1}{3}$ of $\frac{3}{4}$?

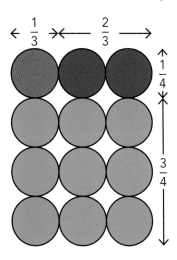

Bar models also help us to visualise what happens when a fraction is multiplied by a fraction. Can you see that:

$\frac{1}{3} \times \frac{1}{4} = \frac{1}{12}$ is the same as $\frac{1}{3}$ of $\frac{1}{4}$?

Dividing a fraction by a whole number

Fraction circles help us to visualise what happens when a fraction is divided by a whole number. Can you see that:

$\frac{1}{5} \div 3 = \frac{1}{15}$ is the same as $\frac{1}{3}$ of $\frac{1}{5}$?

Think about the extra lines you might add to help you see all of the fifteenths.

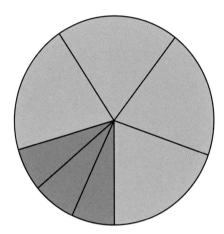

Bar models also help us to visualise what happens when a fraction is divided by a whole number. Can you see that:

$\frac{1}{5} \div 3 = \frac{1}{15}$ is the same as $\frac{1}{3}$ of $\frac{1}{5}$?

Converting fractions and decimals

Remember: a fraction is itself a quotient – the result of a division calculation.

$\frac{1}{4} = 1 \div 4 = 0.25$

```
        0 . 2  5
  4 ) 1 . 0  0
      0 . 8
        . 2  0
        . 2  0
             0
```

Using the column method is one way to find the decimal equivalent of a fraction.

Another way is to find equivalent fractions with a denominator of 10, 100 or 1000.
This makes it easy to convert a fraction to a decimal.

$\frac{1}{4} = \frac{5}{20} = \frac{25}{100} = 0.25$

What is the same and what is different?

0.25 and $\frac{1}{4}$ are two different ways of representing the same number.
They occupy the same position on a number line.

Remember: the order of operations when working with fractions is the same as when working with any other type of number.

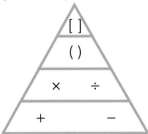

Ratio

A **ratio** shows the relative sizes of two or more values.

	Bar model
The ratio of boys to girls is 1 : 3 (we say 'one to three'). The ratio of girls to boys is 3 : 1 (we say 'three to one').	
Expressed as a fraction: $\frac{1}{4}$ are boys and $\frac{3}{4}$ are girls.	
Expressed as a percentage: 25% are boys and 75% are girls.	

The value of a ratio

For a ratio $a:b$, we call the quotient $\frac{a}{b}$ the **value of the ratio**.

For example:

The ratio of cats to birds is $3:1$. The value of the ratio is $\frac{3}{1}$ (or simply 3). The first quantity is three times the second quantity.

The ratio of birds to cats is $1:3$. The value of the ratio is $\frac{1}{3}$. The first quantity is one third of the second quantity.

The ratio of cars to bicycles is $2:7$. The value of the ratio is $\frac{2}{7}$. The first quantity is two sevenths of the second quantity.

The ratio of bicycles to cars is $7:2$. The value of the ratio is $\frac{7}{2}$ (or 3.5). The first quantity is 3.5 times the second quantity.

Percentage

A value expressed as a fraction of 100 is a percentage. The word 'percentage' means 'of every hundred' and '1 per cent' means 'one part of a hundred: $\frac{1}{100}$'. 100% is the whole.

It is very easy to convert between percentages and fractions:

21% is $\frac{21}{100}$ $\frac{67}{100}$ is 67%

99% is $\frac{99}{100}$ $\frac{14}{100}$ is 14%

50% is $\frac{50}{100}$ $\frac{76}{100}$ is 76%

Proportion

Proportion also describes the relationship between two values.

There is a relationship between the number of chocolate cakes and lemon cakes. The proportion of the plate of cakes that are chocolate is $\frac{3}{8}$.

The proportion of the tray of cakes that are chocolate is also $\frac{3}{8}$.

The **proportion** of chocolate cakes on the plate is the same as the **proportion** of chocolate cakes on the tray.

Proportion

There is a relationship between the width and length of the cabbage patch. The proportion of the width to the length is $\frac{3}{5}$.

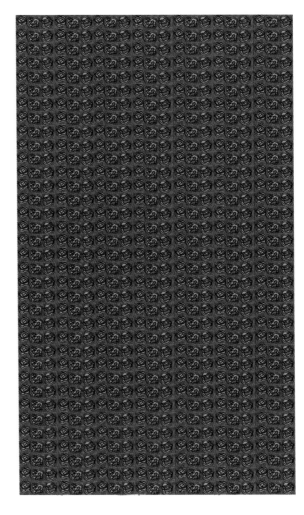

The field and the cabbage patch are in **proportion** with each other because the relationship between the rows and columns of the arrays is the same in both.

Cube

A cube is a 3-D shape with six identical square faces, with three faces meeting at each vertex. Adjoining edges and faces are at right angles.

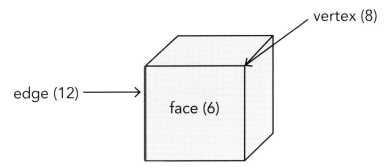

vertex (8)

edge (12)

face (6)

A net is a 2-D representation of a 3-D shape. It is what the shape would look like if it was opened out flat. There are 11 possible nets for a cube.

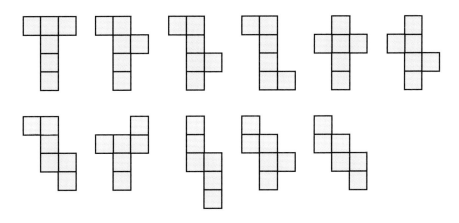

Cuboid

A cuboid is a 3-D shape that is a rectangular prism.

All six faces of a cuboid may be rectangles, or four faces rectangles and two faces squares.

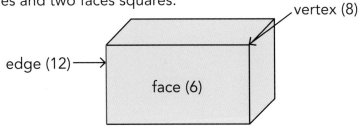

The net of a cuboid is made up of three different pairs of rectangles (or four rectangles and two squares).

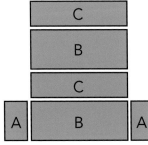

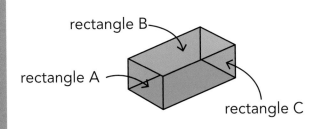

The volume of a cuboid is length × width × height.

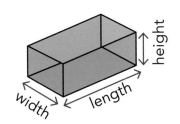

Nets of other 3-D shapes

Triangular pyramid   Square pyramid   Triangular prism   Cone   Cylinder

Reflection of shapes

In a reflection, all the points of a figure are reflected or 'flipped' over a line called the **line of reflection** or **mirror line**.

Every point on the figure moves to a new position that is the same distance from the line of reflection, but on the other side.

The original object is called the **pre-image**, and the reflection is called the **image**.

The image is usually labelled using a prime symbol, for example: *A'B'C'*.

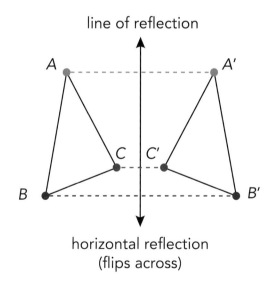

horizontal reflection
(flips across)

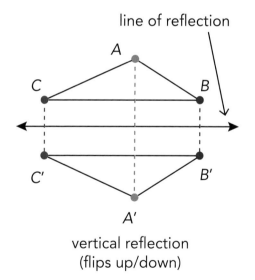

vertical reflection
(flips up/down)

Translation of shapes

In a translation, every point of the object is moved the same distance and in the same direction.

A translation of 2 units right, 2 units down.

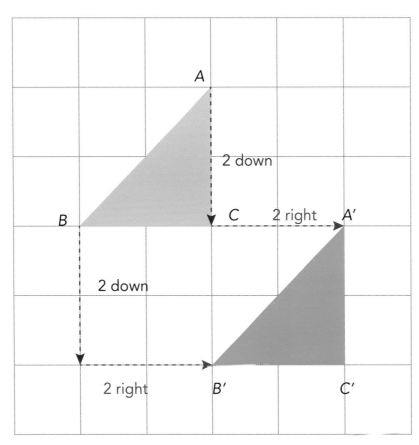

Coordinates and quadrants

The coordinate plane consists of a horizontal axis (the x-axis), a vertical axis (the y-axis) and grid lines that intersect at right angles covering the whole plane.

The intersecting x- and y-axes divide the coordinate plane into four sections. These four sections are called **quadrants**.

Quadrants are named using the Roman numerals I, II, III and IV, beginning with the top right quadrant and moving anti-clockwise. (0, 0)

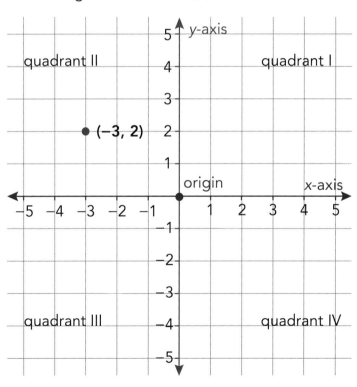

Points on the coordinate plane are described as ordered pairs (x, y) known as **coordinates**.

The location of a point is given by its position along the x-axis (the first value of the ordered pair, the x-coordinate) and along the y-axis (the second value of the ordered pair, the y-coordinate). For example, (−3, 2).

The point at which the two axes intersect is called the **origin**.

The origin is at 0 on the x-axis and 0 on the y-axis (0, 0).

Calculating and solving problems involving the mean

The mean is the most commonly used type of average. To calculate the mean of a set of values, find the sum of the values and divide by the number of values in the set. The mean value will always lie between the highest and the lowest value.

To find the mean of 8, 4, 6, 3, 2 and 7, add the numbers and divide the total by 6 (because there are 6 numbers in the list).

$$8 + 4 + 6 + 3 + 2 + 7 = 30 \qquad 30 \div 6 = 5$$

The mean does not have to equal any of the values. As it is calculated by dividing, the mean is often not a whole number and may need sensible rounding. For example:

Here are the heights of five Year 6 pupils:
140 cm, 149 cm, 138 cm, 152 cm, 143 cm.

To find the mean height to the nearest centimetre:

$$\text{mean} = \frac{(140 + 149 + 138 + 152 + 143)}{5} = 144.4 = 144 \text{ cm to the nearest centimetre.}$$

140 cm   138 cm   143 cm

149 cm   152 cm

Linda joins the group and the mean increases to 146 cm exactly. How tall is Linda?

Here are two ways to calculate the new person's height (H):

$H = (146 \times 6) - (140 + 149 + 138 + 152 + 143) = 154$ cm, or:

$H = 144.4 + [6 \times (146 - 144.4)] = 144.4 + 9.6 = 154$ cm

Tip!

Use algebra to manipulate and solve equations to find the unknown piece of information in problems that involve the mean.

Interpreting and constructing pie charts to represent data

A **pie chart** looks like a pie that has been cut into different-sized slices. The size of the slice shows the relative size of the value.

Pie charts are used to represent data. They cannot show zero or negative values, but they are very good at showing each part's proportion of the whole.

The size of the slices or **sectors** can be shown in degrees, as a fraction, a percentage or just visually. The total of the values is known in each case: 360°, 1, 100% or a complete circle.

Favourite big cats of Year 6 children

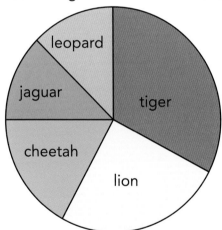

To construct a pie chart, you need to know what fraction of the whole each sector comprises.

From this, you can work out the angle.

Here $\frac{1}{3}$ of the children chose tiger, so the angle is $\frac{1}{3}$ of 360° which is 120°.

With the information given so far, you cannot tell how many children voted 'tiger'. You need to know the total number of children.

For example: If 12 children chose 'lion', there are 12 × 4 = 48 children altogether and 16 ($\frac{1}{3}$) voted 'tiger'.

Could there have been 25 children in the group altogether? (No, because $\frac{1}{3}$ of 25 is not a whole number.)

acute angle: An angle that is smaller than a right angle. It is greater than 0 degrees and less than 90 degrees.

adjacent sides: two sides that meet to create an angle

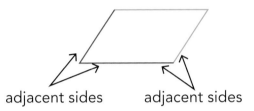

adjacent sides   adjacent sides

algebra: The use of letters to represent numbers in combination with numbers and operators. For example, $3x + 2$ is an expression that uses letters, numbers and operators.

area: the amount of space occupied by a 2-D object

average: a middle or typical value of a dataset

bar chart: a graphical display of data in which the height of each bar relates to quantity or measurement

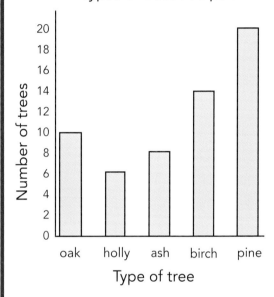

Types of trees in a park

base: the side of a shape that forms a right (90 degree) angle with the height of the object

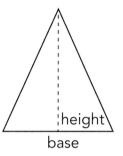

brackets: Symbols used in pairs to group numbers and symbols together; they show which calculation to do first, for example $3(x + 2)$ means 'consider $x + 2$, then multiply it by 3'.

centi: Used in units, it means 'one hundredth' of a unit. So, 1 centimetre is $\frac{1}{100}$ metre and 1 centilitre is $\frac{1}{100}$ litre.

centre (circle): the point equidistant from all points on the circumference of a circle

chord: a straight line segment whose endpoints both lie on the circle circumference

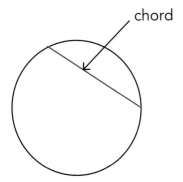

circumference: the edge of a circle

closed cube: a cube with six faces as opposed to an 'open' cube that has one face missing

common denominator: fractions with the same denominator

common factor: A factor that two or more numbers have in common. For example, 6 is a factor that is common to both 12 and 18.

common multiple: A number that is a multiple of two or more numbers. For example, 60 is a multiple that is common to both 3 and 20.

cone: a 3-D shape that tapers smoothly from a flat circular base to a point called the apex or vertex

congruent: Two or more figures that have the same size and shape. The figures are exact copies or exact mirror images of each other.

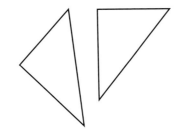

continuous data: Information about something that can be shown on a continuous scale; measured rather than counted, for example length or mass.

convert: change to something equivalent

$$\frac{1}{2} = 0.5$$

$$8:4 = 4:2 = 2:1$$

$$1000\,\text{m} = 1\,\text{km}$$

coordinate grid or **coordinate plane:** a grid used to locate points in space

coordinates: the two numbers that identify the position of a point on a coordinate grid or coordinate plane

cuboid: a 3-D shape with 6 rectangular faces (or 4 rectangular and 2 square faces), 12 edges and 8 vertices

cylinder: a 3-D shape that has two circular bases joined by one continuous curved surface; the cross-section of the shape is uniform and a circle

data: information that has been collected and can be compared

decimal fraction: a fraction with a denominator of 10, 100, 1000, …

decimal point: the symbol (a dot) that separates the whole number part from the decimal part

deconstruct (shape): split a shape into smaller, component shapes

denominator: The bottom number in a fraction. This shows the number of equal parts the whole has been split into.

$$\frac{7}{12} \longleftarrow \text{denominator}$$

diagonal: a straight line that goes from one corner to an opposite corner

diameter: any line that joins two points on the circumference of a circle and passes through the centre

discount: a reduction from the original value, or price

discrete data: data that is counted – items that exist separately from each other

dividend: the whole quantity before it is divided

division: An operation where something is divided; a number is split into equal parts or groups using the ÷ symbol. Fractions are a way to represent division calculations, for example, $\frac{10}{5}$ is the same as 10 ÷ 5.

divisor: the number by which another number is divided

$$136 ÷ 8 = 17$$
$$\uparrow$$
$$\text{divisor}$$

equation: a mathematical statement containing an equals sign, for example 3x + 2 = 8

equilateral triangle: a triangle in which all three sides are equal (and therefore all angles are equal)

equivalent: equal in quantity, size or value

equivalent fractions: fractions that have the same value

$$\frac{3}{5}$$

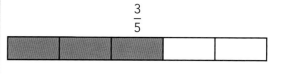

$$\frac{6}{10}$$

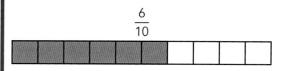

equivalent ratio: ratios where the value is the same, for example 1:2 and 3:6 both have a value of $\frac{1}{2}$ $(= \frac{3}{6})$

evaluation: to find the value of a mathematical statement

expansion (of brackets): to remove the brackets, evaluating the mathematical statement

$$3(a + 2) = 3 × a + 3 × 2 = 3a + 6$$

expression: a combination of numbers, symbols and operations arranged to show the value of something, for example 3x + 2 is an expression

face: the flat surface of a 3-D shape, bounded by edges

factor: a number that divides exactly into another number with no remainder

formulae: mathematical relationships or rules written using symbols, for example the formula for the area of a rectangle is $A = l \times w$, where A = area, l = length and w = width

fraction: part of a whole, shown by writing one number (the numerator) on top of another (the denominator)

height: the perpendicular distance from the base to an opposite vertex or side

image: the new position of a point, a line, a line segment or a figure after a transformation

improper fraction: a fraction that is greater than one; where the numerator is greater than the denominator

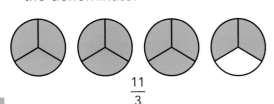

$$\frac{11}{3}$$

inequality: a mathematical statement showing unequal values or expressions, for example $3x < 4$ is an inequality, which states that $3x$ is less than 4

integer: a whole number; a number without a fractional part

intersection: the point at which two or more lines meet or cross

isosceles triangle: a triangle with at least two equal sides

kilo: Used in units, it means 'one thousand' of a unit. So, 1 kilometre is 1000 metres and 1 kilogram is 1000 grams.

lateral face: any face of a 3-D shape that is not a base

line graph: a graph where connected data are plotted and connected by lines to show how something changes, usually over time

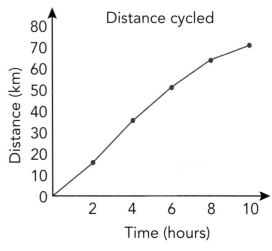

line of symmetry: a line through a shape that can be imagined or drawn which divides it into two parts that, when folded onto each other, will cover exactly the same shape and area

line segment: a straight line that connects two points

A ————————— B

linear (number sequence): a sequence generated by adding or subtracting the same amount each time

$$2, 5, 8, 11 \dots$$

loss: In the context of pricing, loss describes the amount of money lost after deducting the original buying cost of the item when selling.

lowest common denominator: the lowest value of all the common denominators in a set of fractions

mean: the most common type of average; calculated by dividing the sum of the values by the number of values in the set

milli: Used in units it means 'one thousandth' of a unit.
So, 1 millimetre is $\dfrac{1}{1000}$ metre and 1 millilitre is $\dfrac{1}{1000}$ litre.

mirror image: the reflection of a point, a line, a line segment or a figure as seen in a mirror

mixed number: a number that consists of a whole number and a fraction

$$14\tfrac{1}{3}$$

natural numbers: positive numbers from 1

$$1, 2, 3, 4, 5, 6 \dots$$

net: a 2-D figure composed of polygons which could be folded and joined to form a 3-D shape

net of a cube

net of a tetrahedron

nth term: An expression that enables us to calculate the term that is in the *n*th position of a sequence. *n* stands for any position number. $3n - 1$ is the expression that describes the sequence that starts 2, 5, 8, 11, … This means that in each position, the value is 3 times the position number, then subtract 1. For example, in position four, the value will be $3 \times 4 - 1$ (11).

number sequence: a series of numbers that follow a pattern

Sequence for $3n - 1$

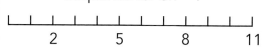

numerator: The top number in a fraction. This shows the number of equal parts out of the whole that are being considered

$$\dfrac{3}{4} \longleftarrow \text{numerator}$$

obtuse angle: An angle that is bigger than a right angle but smaller than a straight angle. It is greater than 90 degrees but less than 180 degrees.

opposite sides: two sides of a shape that have no common vertex

ordered pair: Two numbers for which the order in which they are given is important. Coordinates are presented as ordered pairs.

origin: the point on the coordinate plane at which the x- and y-axis intersect, where both x and y = 0

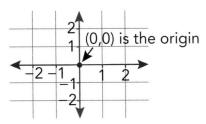

(0,0) is the origin

parallel lines: lines that stay the same distance apart along their whole length; they will never meet

parallelogram: a quadrilateral whose opposite sides are parallel and equal in length

percentage: a number expressed as a fraction of 100

$\frac{1}{100}$ is 1%

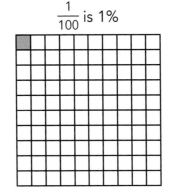

$\frac{16}{100}$ is 16%

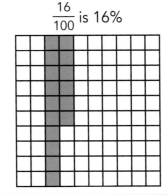

perpendicular: a line or plane that is at right angles to another line or plane

perpendicular foot: the intersection of the base and perpendicular height of a shape

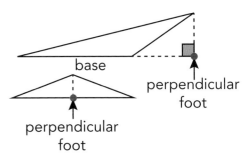

perpendicular lines: lines that are at right angles to each other

pie chart: a circular chart divided into sectors where the different sectors show the relative size of each value

Children's favourite vegetable

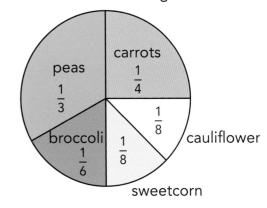

prism: a 3-D shape with two parallel and congruent polygonal bases named after the shape of the base

 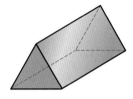

hexagonal   triangular
prism   prism

---

profit: the difference between the cost of an item (to buy or make) and the price the seller sells it for

---

proportion: the comparative relationship between a part and the whole that it is part of

---

protractor: an instrument for measuring angles

---

pyramid: a 3-D shape with a polygon base and triangular sides which meet at a single point (apex)

quadrant: One of the quarters of the plane of the coordinate plane. Quadrants are numbered I, II, III and IV, going anti-clockwise from top right.

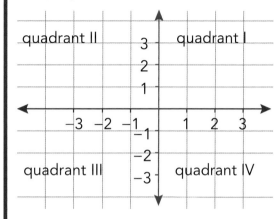

---

quotient: dividend ÷ divisor = quotient

$$430 \div 2 = \underline{215} \longleftarrow \text{quotient}$$

---

radius (radii): the distance from any point on the circumference of a circle to the centre

---

range: the difference between the highest and lowest values in a data set

ratio: Shows the relative size of two or more values. In the example below, the ratio of triangles to circles is 3 : 5. This means that, in the group, for every 3 triangles there are 5 circles.

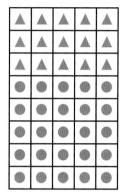

---

recurring decimal: a decimal number in which a digit or a sequence of digits in the decimal part repeats forever

52.67676767...

reflection: a transformation that moves a figure to its mirror image by 'flipping' it over a line of reflection

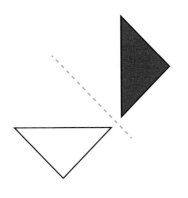

right angle: an angle of exactly 90° (degrees), corresponding to a quarter turn

round off / round: Rounding off (same as just 'rounding') means approximating a value in terms of the whole (or tenth, or hundredth and so on) **nearest** to it on the number line. A judgement must be made whether this is the value above or below. For example, to round 3.17 to the nearest tenth we must express it as 3.2. Any value that is exactly half way between two values is always rounded up to the next. So, 4.15, rounded (or 'rounded off' or 'rounded to the nearest') to the nearest tenth must be rounded to 4.2.

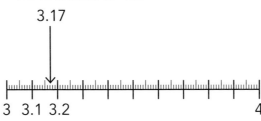

3.17 rounded (or 'rounded off') to the nearest tenth is 3.2.

round down: Rounding down means approximating a value downwards by expressing it in terms of the whole (or tenth, or hundredth and so on) below it. For example, to round 3.17 down to the nearest tenth below 3.17 we must consider it as 3.1.

3.17 rounded down to the nearest tenth below is 3.1.

round up: Rounding up means approximating a value upwards by expressing it in terms of the next whole (or tenth or hundredth, etc.), for example, 3.17 rounded up to the next tenth is 3.2

scale (on a map or a drawing): the ratio of the distance on the map (or drawing) to the actual distance on the ground (or real life object)

scale factor: The multiplier that shows how one value relates to another; often used to describe relationships between lengths, areas or volumes of similar shapes.

A is multiplied by scale factor 4 to become B. This means that values for B are 4 times those for A.

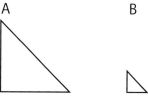

A is multiplied by scale factor $\frac{1}{3}$ to become B. This means that values for B are $\frac{1}{3}$ times those for A.

scalene triangle: a triangle that has three unequal sides

sector: a 'pie-slice' of a circle; the area of a circle covered between two radii

sequence: An ordered list of numbers or objects arranged according to a rule. For example: 2, 5, 8, 11, … is a sequence. The rule is that each number is 3 more than the previous number.

similar shapes: Shapes that are identical in shape but not in size. Their sides are in the same proportion.

simplest form: A fraction is in its simplest form when the numerator and denominator cannot be simplified any further. No smaller numerator or denominator can be found that will give an equivalent fraction of the same value, for example $\frac{7}{21}$ in its simplest form is $\frac{1}{3}$.

simplest form of a ratio: the whole number form of a ratio with no common factors

simplify: To reduce the numerator and denominator in a fraction, whilst ensuring it keeps the same value. Fractions are simplified by finding common factors between the numerator and denominator.

solution (in algebra): A value that is put in place of (or 'substituted for') an unknown to make the equation true. For example, in $2x + 7 = 13$, by substituting 3 for x, the equation is true, therefore, the solution of the equation is 3.

solve (in algebra): work out the value of the unknown(s)

symmetry, symmetrical: A figure is said to be symmetrical if it can be reflected or folded across a line so that one part fits exactly on top of or is the mirror image of the other part.

term (in sequences): a number in a sequence, for example in 2, 5, 8, 11, …, the numbers 2, 5, 8 and 11 are all terms in the sequence

term-to-term rule: The operation carried out on a number to create the next number in a sequence. For example, in 5, 9, 13, 17, …, the term-to-term rule that generates this sequence is + 4.

translate: to move every point of a figure by the same amount left or right and up or down

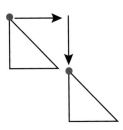

unknown: a number the value of which is not known, for example in $3e + 4f = 12$, the e and f are unknowns

value of a ratio: given a ratio $a:b$, we call the quotient $\left(\frac{a}{b}\right)$ the value of the ratio

variable: a symbol used to represent an unknown number, for example in $2d + 4 = 11$, where d is a variable

vertex (vertices): a point where two or more straight lines meet

x-axis: the horizontal axis of the coordinate plane

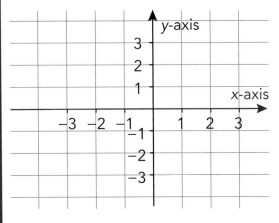

y-axis: the vertical axis of the coordinate plane

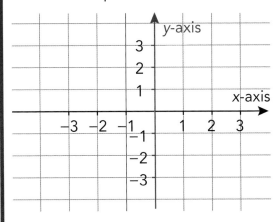

1	2	3	4	5	6	7	8	9	10
11	12	13	14	15	16	17	18	19	20
21	22	23	24	25	26	27	28	29	30
31	32	33	34	35	36	37	38	39	40
41	42	43	44	45	46	47	48	49	50
51	52	53	54	55	56	57	58	59	60
61	62	63	64	65	66	67	68	69	70
71	72	73	74	75	76	77	78	79	80
81	82	83	84	85	86	87	88	89	90
91	92	93	94	95	96	97	98	99	100

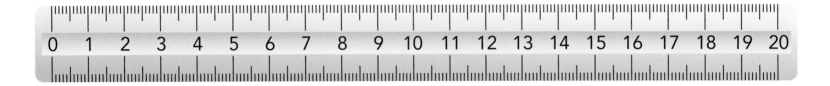

1	5	10	50	100
I	V	X	L	C

4	6	9	11	40	60	90
IV	VI	IX	XI	XL	LX	XC

18	34	59	92
XVIII	XXXIV	LIX	XCII